The Psychology of Space Exploration

This short book grapples with two vast questions: the nature of our minds, and our place in the wider universe. It considers how one mutually influences the development of the other.

The changes and challenges that will accompany the first humans to leave Earth and travel to another planet, or even further, will not only impact our technical capabilities, but will also represent a watershed moment within our individual and collective human psychology. Many of the problems of resource use, environmental degradation, and waste or destructive processes are contained in the larger process of exploring another environment and planet. This book also offers a shift in perspective that allows us to consider humanity from an alternative, more holistic perspective, reappraising our own minds both individually and within dynamic social processes.

The Psychology of Space Exploration considers our place and purpose in the widest possible perspective, that of space exploration and the natural universe. It doesn't seek to answer these questions, but provides a perspective to explore even further.

Dr. Richard Sherry is a Consultant Chartered Clinical Psychologist HCPC Reg. and BPS CPsychol, AFBPsS, a BPC Reg. Psychoanalytic Psychotherapist, and an APECS licensed Executive Coach specialising in elite leadership and high performance. He is a specialist in psychological clinical traumatology (ESTSS Cert and an EMDR Consultant). Much of his lifetime's work is in understanding how to reverse cycles of trauma within complex and high stress environments. His work in psychologically treating very high performing individuals/ teams/organisations as well as vulnerable populations and complex presentations has taken him around the world. Much of this expertise in working to help treat and heal individuals, groups, and organisations. Much of his work has been looking at ways to help address social and emotional trauma at every level to look at how we can heal these difficulties to develop higher aspects of healthy and compassionate lives and consciousness of the world.

In addition to some of these previous details, Dr. Sherry has worked in posts as the psychologist within the fire service as well as the National Service for Police Psychiatry, and he has headed the Clinical Psychology section for the US Military inpatients treatment for more than seven years. He has specialist training in clinical neuropsychology, holds a Diploma in post-graduate Ethics with a specialism in Disaster Ethics, has an MA from the Tavistock Clinic in Organisational Consultancy, and completed specialty training in Executive Coaching Psychology where much of his focus is on innovating and evolving beyond our blocks. Dr. Sherry also has specialty training in extreme medicine; he is working to finish his Fellowship in Wilderness Medicine (FAWM), where his passion is the interface of leadership within areas such as complex disasters as well as the psychology of deep space exploration and travel. Much of his work has focused on the pivotal understanding of innovations in low and zero space environments and how these challenging environments can be transformed through profound perspective shifts in technology and human capital.

The Psychology of Space Exploration
What Freud Might Have Said

Dr. Richard Sherry

Routledge
Taylor & Francis Group
LONDON AND NEW YORK

First published 2019
by Routledge
2 Park Square, Milton Park, Abingdon, Oxon OX14 4RN

and by Routledge
52 Vanderbilt Avenue, New York, NY 10017

Routledge is an imprint of the Taylor & Francis Group, an informa business

British Library Cataloguing-in-Publication Data
A catalogue record for this book is available from the British Library

Library of Congress Cataloging-in-Publication Data
A catalog record for this book has been requested

ISBN: 978-1-138-35142-4 (hbk)
ISBN: 978-1-138-35140-0 (pbk)
ISBN: 978-0-429-43528-7 (ebk)

Typeset in Bembo
by Apex CoVantage, LLC

MIX
Paper from responsible sources
FSC® C013056

Printed and bound in Great Britain by
TJ International Ltd, Padstow, Cornwall

To my parents, Mark, Mike, Bob, and Brett
and most importantly my two wonderful sons
thank you all for encouraging my deep creativity and stellar aspirations.
Sharing the right conversations can help us change the world—

Contents

1 Introduction

Early in the evening of spring 1995, I was walking up the crest of the hill overlooking the lakeside. All of a sudden I came upon the largest super moon I could have imagined. It hovered before me, seemingly sitting upon the trees on the other side of the lakeshore. I stopped and spent nearly half an hour just staring in complete awe at this glowing sphere of light.

My brain reinterpreted the perceived distance, making the moon appear so colossal that, if I reached out my hand far enough, I felt I would be able to brush against its dry and dusty surface. The illusion was so dramatic that these two different worlds appeared to come together, one almost seemingly colliding with the other. Something about this experience changed me. I have never been able to describe how, but the circumstances, and this encounter, redefined my horizons.

This experience occurred during the same period that I was taking my first class on Freud at university. I can date both of the two central themes of when my interest in space (or at least in the greater outer world) as well as my life-long passion with Freud began. The awareness of this extraordinary and unknown external world seemed to equally open the door to a correspondingly breath-taking inner universe. It was these two equivalent three-dimensional playgrounds of space that ignited my own imagination and inner life that these places beckoned to be explored and understood. Much of my life has been connected with exploring and understanding how these potential environments can be understood and developed.

If we return to the central question, why should you as the reader be interested in these areas I have been working to understand, such as space psychology; or why would you take any interest in a thinker from 100 years ago? What relevance is space or the psychological aspects of

***Figure 1.1* Photo of Super moon**

Source: www.pexels.com/photo/white-and-black-moon-with-black-skies-and-body-of-water-photography-during-night-time-748626/

these issues to my everyday life or to me? All of these are valid points, but as I hope to illustrate, there is a subtle, but profound relationship to a much wider and deeper world that most people are not aware of and which does affect their everyday lives. The quality of our lives and even potentially to our very survival is predicated on properly resolving these questions. Not to put too fine a point on it, but getting these aspects right could very well determine the future of human beings on this planet or any others.

2

Introduction II

Having had a tour of some of the key aspects and processes within our own internal worlds as well as having a clearer orientation to our cosmological context, we need to integrate the uniqueness of the time period that we are now inhabiting. When I was a child, the year 2017 for some reason seemed to be equated with the future. I am now 43 years old. Similar to my earlier sense, I feel that we are facing a complexity of change with the potential of irreversible impacts on a global and planetary level that really does require entirely new levels of integration of multi-disciplinary thinking. Issues like global warming, the extent of urbanisation, and the mismanagement of natural energy resources that could accelerate changes within our biosphere and our air, soil, and water quality have been described as 'ecocide' (Higgins, 2015). Why would this topic be of any interest to a psychologist and psychoanalyst?

The easiest answer is: what does one study if not consciousness and the practice and existential interaction of the study of minds in concert with their environments?

The first layer of understanding the importance of space travel is in its essence about properly connecting with our curiosity. What is on the other side? For the first sea explorers, the uncertainty and corresponding feelings of fear were probably equal factors to the risks and potential rewards of leaving everything behind to venture into new worlds to discover and conquer. Part of human psychology is to dream and to go experience different places. Most of the representations of the beginning literary and cinematic portraits of space travel and exploration contain an extraordinary intensity of anxiety and foreboding, if not the development of complete psychological breakdowns. Why might this consistent portrayal of psychological collapse and space exploration be the benchmark of the artistic imagination? These kinds of pessimistic cinematic and literary versions influence how the public sees the science fiction.

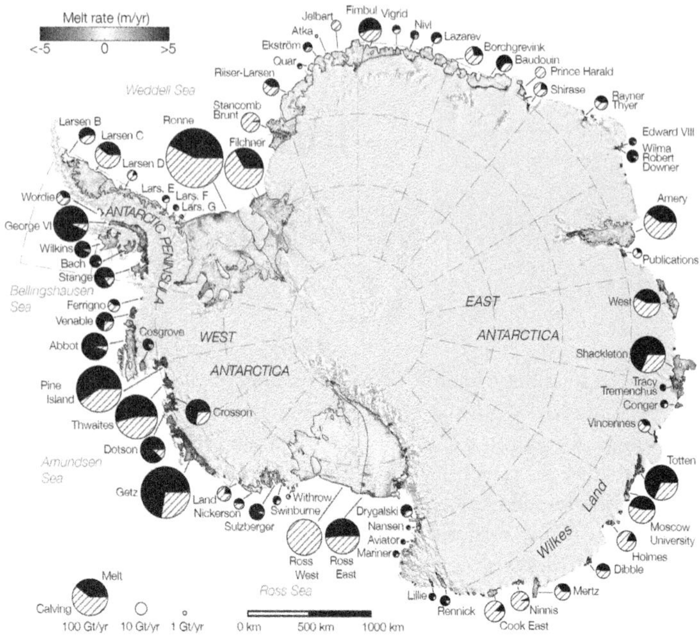

Figure 2.1 **NASA melt map of Antarctica**

Source: www.nasa.gov/images/content/755402main_earth20130613b-full.jpg

It is possible the predominant views could forestall mission success sealing our collective scientific imagination in creating the ingredients for a successful mission (or not) into space? Technically, the severity of psychological anxiety states related to the deep unknown portrayed contain within them the ridding of the emotional states of panic and terror. These uncertainties are evacuated from the depths of the person's internal world. These kinds of fears about the unknown external world then become projected to the external world in an attempt to have these gotten rid of. In the cinematic versions of space exploration, frequently these paranoid anxieties are expressed as embodied into the other team members, the living milieu, or the surrounding environments (spacecrafts or planetary environs). Looking at these questions about the unknown necessitates that we think about their implications in potentially new ways to take a step back from these processes to better appreciate how anxiety works, especially within the face of the absolute limits of space.

3

Freud's impact on my view of human nature

In the very early part of my career, I spent nearly two years working with Professor Mark Solms, when he was living in London, as his researcher helping him with the *Standard Edition of Sigmund Freud's Complete Psychological Works* (24 volumes). Mark Solms is the father of neuro-psychoanalysis, and is an eminent psychoanalyst and Chair of Neuropsychology now in Cape Town, South Africa.

During this time, I worked on Freud's complete body of psychological writings, reading through it twice and focusing on making updates to help modernise this work for the new revised edition. More precisely for this period of my life, I layered my day by starting off working either with new parents and neonates in the obstetrics hospital, while alternating with working at the UCL children's crèche for my jobs that connected with my post-graduate training in early child development. I was undertaking the child development course of the first part of Tavistock's psychotherapy training more years ago than I care to remember. After this was finished, I would take the unprecedented step of being one of the first Tavistock trainees to go over to the Anna Freud Centre (which is only down the street from Tavistock). At this point, however, because of a (perceived differences in technique between Klienian and Anna Freudian schools of psychotherapy) a huge rift in communication between the two groups was previously present between the Kleiniens' (from the Tavistock) and the developmentalists from Anna Freud that is now beginning to thaw. In this way my interconnectivity going between each school was completely unique. After I finished at the Tavistock, I would go to the 'Wednesday Anna Freud Child Analytic Lectures' for the rest of the week; I would go straight to the Anna Freud Centre to do my evening work, and made myself a large pot of coffee to help with the late evenings of reading and editing my way through Freud's collected works. By design and some help and support from my mentors, these experiences helped me start off the day deeply immersed in the psychological realities of early real life and trying to take these everyday human social interactions, like seeing new parents with their neonates (and I saw probably more than 2,000 parents) begin to make sense of having their own baby. This carried over into my other work environment where I had research opportunities every day, like hearing a three year old's joke and studying the practical aspects of children's social lives and play. By the evening I would bicycle up to the Anna Freud Clinic to work through all of Freud's *Standard Edition*, and where I would work late into the evening.

These experiences provided a very deep appreciation in the subtlety of the developmental emotional lives of human beings through some of the worst difficulties. For instance, dealing with the impact of still births, psychotic postpartum depression, and stresses at home for the nursery children, to seeing when every aspect lined up beautifully and what differences this could make. These experiences really gave me a profound appreciation for our internal and emotional lives (even intergenerationally) that has impacted the rest of my professional and personal life thereafter. There is a magnificence, depth, and beauty to a full human being and a life that I still have never felt has yet been captured fully in literature or in art. I feel it is so much greater than I have yet to see it portrayed.

I felt so fortunate to have these different and really profound experiences. Every one of these layered experiences I have felt moved me and helped me grow into a deeper and more compassionate human being. Connected with reading Freud's writing I have lived, breathed, and had Freud's work saturate my being several times over. At least for this part of his writing, I think it is helpful to think one of the genius qualities of his own thinking was that he tried at some level to think about the entirety of his psychological works as part of a systematic examination of some of the biggest and most challenging questions human beings might be able to ask ourselves. However much of an achievement it was to read this part of his corpus, keep in mind that Freud wrote all of this work, plus his neuropsychological papers, and his correspondence which are actually from a corpus of work that is greater than all of the other literary output combined together. This really is a terrific accomplishment in volume, not to consider similarly how revolutionary some of his key words, seminal concepts, and thinking have been on every aspect of contemporary thought including writing and socialising with some of the greatest scientists and artists of history from his own time, which continues directly up to our own era.

Many have considered his book *The Interpretation of Dreams* to be his magnum opus. Within this volume, he explores the logic and unfathomable boundless architectures of human experience, memory, and the interaction of these with the instinct and psychical drives within the hidden world of sleep. All of these aspects of human life express the unconscious within the tapestry of interacting with the imprint of our social universe, which he described like a wax scribe indenting into our memory an imperfect recording our own subjective perspective of our everyday lives. These ingredients of human biology can help us better understand our human nature, especially as these influence our dreaming lives as well as our imaginations, remapping onto one and other to draw out themes and truths that seem to go beyond what we are consciously aware of. Each one of us has a direct and nightly relationship to these forces, whether we realise it or not, that offers a similar limitless scope of creativity that actually does impact on us even if we are unaware of it. This core experience of deep, hot imagination really is something unique and special that is driven by the frameworks of the neurobiology of our emotional systems and our need to sate meaningful interactions with others in a social world. It is humbling to think that each of us has a direct and unfolding relationship with this over our entire lives. There is a phrase from Hemmingway, where he describes 'a moveable feast.' I think that this richness of our emotional experience and what can be present in and over the course of our lives is nothing short of miraculous.

Figure 3.1 **NASA photo of Paris at night**

Source: www.nasa.gov/sites/default/files/thumbnails/image/iss043e093480_lrg.jpg

4

Connecting the necessity of the outer world to our inner universe

Even on a deeper and more subtle level, as you read through this book, I would like to create a profound awakening for you to internalise and build an appreciation that I have come to love in engaging with both of these inner and outer spaces. There are many things we do not really give much thought to. For example, the cell phone or Internet that you probably use so many times a day runs off satellites. Both of those technologies were developed via space innovation, and moreover they remain working only because of these sets of specific expertise. These same ubiquitous everyday technological necessities can just as easily disappear with the equivalent of a 16-car pile-up in near Earth orbit. This gravitational level of Low Earth Orbit (or LEO) is where the satellites and the International Space Station (ISS) are in continuous flight path. The distance from the Earth is roughly between 198.6 and 215.5 miles above the Earth's surface (Space Foundation, 2017) and because of the orbital velocity, if one of these objects hits another one it could set off a balletic set of collisions that would be the equivalent of thousands of 50 calibre machine-gun rounds slicing through everything the debris touches. The result could end with a catastrophic accumulation of more and more space debris, thus creating a technology-altering event. This small amount of space debris could potentially, in half an hour, end what we have become accustomed to living with in our use and relationship with modern technology. Another challenging scenario comes from the sun's coronal mass ejection, a solar flare. If it was big enough, it could send an electromagnetic pulse (EMP) sufficient enough to fry any exposed computer and undermine our electrical systems as well. This could potentially send our digital age back to something at least reminiscent of the Stone Age, as if we had a huge magnet collectively wipe clean all of our stores of coveted daily

Figure 4.1 **NASA photo of ISS hovering above the Earth**

Source: www.nasa.gov/sites/default/files/images/557343main_iss027e036700_full.jpg

peta-bytes of data. All of this electronic knowledge potentially could be gone. Our relationship to these wider networks of modernity does need to be understood and we need to appreciate how these experiences also change our consciousness in certain kinds of ways that we may not have given thought to.

These are just two points in a myriad of many surprising, shocking, and joyous cosmos of ideas that I would like you to accompany me on in this 25-year quest I have been undertaking thus far to grapple in this very short and digestible book to open the conversation about these inner and greater outer extraordinary spaces. In many ways this writing is a tremendously personal narrative connecting many of my deepest hopes, fears, dreams, and aspirations that I would like to share with you in my journey to grapple with these unique places and spaces. I have been deeply struck: the more that I have immersed myself into the topic of the two genuinely mysterious spaces of outer space as well as the unconscious inner space, the more beauty and depth emerge much more clearly.

Both of these journeys have deepened my enjoyment and my sense of living a meaningful life in addition to creating a growing sense of awe and gratitude of how the world works as well as it does. Each of these spaces has a dynamic quality of energy, beauty, and movement of which we can properly appreciate and create an understanding of. Each of these realms will profoundly enrich our lives – especially in connecting us to a greater sense of gratitude and appreciation for how to nurture an ownership of responsibility to care for these gifts we have been enfranchised to look after.

5

Understanding the context of outer space and of the inner space of the unconscious

Tour of our inner world and that of our outer universe

Internal space

There are different ways we can think about life and also the qualities of 'aliveness.' This definition would clearly change in the discipline of space science by asking if a virus on Mars is alive? There is a quantum jump in complexity to using the conceptual approaches from my own profession. As a clinical psychologist, a specialist in neuropsychology, and a psychoanalytic psychotherapist, I have spent my career understanding the mental life of human beings and social systems – usually under extreme stress. This work has stretched and evolved from looking at and treating the worst kinds of psychological trauma to developing high performance levels with coaching. There is a key relationship between the worst of humanity and seeing the best we are capable of creating.

The specifics of and the clear qualities of our subjective reality underpin much of the foundations of both why psychological science is complicated, just as much why it is relevant to understanding so much of how we put the world together. Within these parameters I would approach this precise question of 'aliveness' in a very different way than most other professions. Just because, for example, a person is physically alive, they can breathe, eat, etc., this does not mean that their emotional levels and qualities of aliveness would necessarily be complete. For instance, it is not just the person's absence of depression that needs to be looked at, but what are his or her positive qualities, like emotional warmth, engagement, or even more importantly, love? Can the person experience the joy of this feeling and the connection it brings? Can he or she accept and express this kind of connected attachment? We equally need to understand the importance of these qualities as the personality is made up of an interactive layering of experiences that activate and synergise these alive relationships and connections with others. For instance, the unique personality of the person continues to layer on to other evolving life experiences. How the individual makes sense of these emotionally salient experiences all integrates to help them make sense of their world. Furthermore, the person, to make meaningful changes in the world around them, uses these belief systems. These factors are all critical parts that layer up over time to help create the person's internal world.

With the internal world we move into the *felt sense of meaning* and the *internal relationship of understanding* that emerges from these events. A human being can have a traumatic event that may or may not impact upon the nature of their character and their expectations of what it means for them to live in the world. With difficult experiences, a person's subjective response to this can change the sense of joy or meaning that he or she can feel in any number of different ways. When the event(s) does impinge upon the person, he or she can feel frequently overwhelmed by an engrained sense of misery, and this emotional 'aliveness' creates a feeling where they are subsisting rather than creatively thriving. I would say that these are some of the reasons why, in my professional opinion, the emotional extremes that understandably would be magnified within environments such as space is spectacularly special. These emotions, and how we think about them, now make the precision of how personality, experience, presence, subjective reality, vision, and the applied leadership to bring all of these together to create something is very hard to quantify in the same way we might be able to calculate a problem such as with astrophysics. Sir Isaac Newton, when asked what he thought of the infatuation of the people, answered that he could calculate the motions of erratic celestial bodies, but not the madness of a multitude (see endnote).

This point highlights the enormous courage required to really try to make sense of the invisible complexities in the emotional life of human beings. In this way, Freud's thinking attempted to systematically grapple with what Newton could not figure out, namely how to calculate the 'madness' of people. What I am interested in is understanding the quality and the depth of that emotional experience that links to and unlocks deeper capacities to create paradigm-shifting experiences. This requires a self-conceptualisation of understanding our own mind in concert with dynamic relationships with other minds and presences

Figure 5.1 **Astronaut on the ISS with floating oranges and lemons**
Source: www.nasa.gov/sites/default/files/thumbnails/image/20897757635_8561e92d60_o.jpg

and technologies. In this way, more fully conceptualising healthy emotional relationships takes this intricacy to an entirely new level.

Like an artist learning to draw or paint by copying from the Master's greatest works of art, we need to struggle with profound and worthy problems using some of the best thinkers to hone our capacity for thinking as well as strengthening the imagination to find extraordinary solutions. I very much feel that Freud holds this level of genius within his thinking, which is made accessible through his writing.

Space travel

There is an intertwining process of development between human imagination and the capacity to do something entirely new. For example, voyage to a different planet, where all of these practical innovations, requires the spark of imagination to work as the catalyst for these kinds of worthy changes. Similarly, the prospect of space travel and the tipping point to activate the physical movement toward this potentially new direction in human capability will change the fundamental ways we move forward as a species and as a planet. In each of these theatres of life we can plumb and try to make sense of the dynamic interrelationships of these influences and try with much greater insight to correctly work to take compassionate ownership of how our (individual as well as collective) actions play themselves out to influence the greater quantum of stability and health.

Like with my own work, returning to these greater thinking spaces and puzzles does sharpen the mind to contend with grappling with these cryptographic problems that connect up material reality with that of the 'hard problem' of what creates the spark of consciousness and sentience. These are very worthy queries to grapple with. It is part of our make up for us to be curious about whom we are in relationship to everything else. These uncertainties connect Goethe's red thread of Nature, with that of the pull and push of the life force within us that contains so much mystery and wonder, just as with feelings of fear and conflict. The theoretical physicist, Professor Albert Einstein, has noted that the greater the value truth holds within a concept also confers that these ideas contain elegance and beauty. This in turn, the elevated quality of exquisiteness, has a greater probability of correctly describing the nature of truth contained within it.

For my life thus far, I have been deeply privileged to study arguably just about every main aspect about the human brain, mind, consciousness, and its resultant output with the worst expressions of human violence to the highest and most tender expressions of human kindness. When I was growing up, I would constantly go into work with my father who was a professor of neuroradiology (he is now working in private practice). He is a medical doctor of the brain and the spine. He reads magnetic resonance images (MRI's) and also performs very highly skilled interventions in the brain and the spinal chord with an array of guide wires and microscopic tools for interventional micro vascular neuro-surgery. With his help and tutelage, I learned to love all of the different aspects of the neuroanatomy of the human brain, such as the diversity within the different parts of the brain. This lifelong and continuing passion in exploring the brain continued with the monumental work of Professor Jaak Panksepp, who developed the field of Affective Neuroscience. Affective Neuroscience, for example, studies the areas responsible for feeling pleasure (the nuclear acumbans); panic or fear are some of the other areas within these neural systems. Other focal areas like seeing and recognising a human face (fusiform gyrus), helping us remember (the hippocampus), and even feeling our body (the insular cortex) are just a few of these areas of brain localisation.

The gifted neuroscientist Professor Bud Craig has described that *the insular cortex* works on so many extraordinary levels that it *reassembles our experience of time and space together*, thus literally assembling our physical and emotional experience of time in what he has called "the global emotional moment" (Craig, 2009). This is a very new way to think about the neuropsychology of how we are put together. If we, like physicists, can update the models of explanation, we may be able to more accurately describe as well as treat illnesses successfully.

I have looked at these different parts of the brain and body under histological slide microscopes. I have also worked with neurological patients throughout my training, seeing what happens when specific parts of their brain do not work the way they are supposed to. A good deal of my current work with individual patients is working with them when there are particularly high levels of distortions that affect their everyday functioning with how they feel about themselves (their identity) with body dysmorphic disorder (BDD) and also somatic problems with psychical unexplained symptoms. I have worked with many BDD patients that would spend hours looking at their skin (particularly in the mirror) and picking at it. Or I have had many patients that had a spectrum of epilepsy, including being able to create seizure-like responses in their brain that would cripple them, causing them to lose consciousness. These particular mental health difficulties create specific alterations that can literally transform the health of the person to where their well-being is spectacularly undermined and feelings of profound vulnerability manifest. Knowing how all of these different levels line up together can really help create understanding and specifically separate out the highly complex interaction between the physical and the emotional. Moreover, it can also highlight what these two do when they do not have a relating overlap. My point is that our internal emotional worlds can have a potential presence and impact much like a hurricane can on the Earth, such as a galactic planetary impact can influence the physics and gravity of the surrounding celestial bodies. Our moods and behaviour can be drastically changed through how we perceive reality and experience these issues. It is whether we are able to describe these perceptual shifts it as clearly in these kinds of ways with our current technology and descriptions.

6

Space exploration in the tension between cooperation and sublimation

Finding collective points of working through cultural anxieties

This is why my entire previous corpus of research is being utilised in these grandest of problems. Space and our relationship to the wider universe and our place within it has come to captivate me. If we are stuck within our own self-serving ego without being able to pick up our head to behold the potential vastness of the blanket of stars and the presence or absence of life that is contained within it, we have no way to go beyond our own insulating tendencies. For me this is *the* recipe for ignorance. In its essence, what makes us human is our capacity to relate and to contend with the value and meaning of these greater experiences. However, like the value of travel, which originates in the French verb *travailler*, which means to work, these kinds of experiences are by their very nature tough, and can be harsh and unpleasant. Any new development in the innovation of venturing to new places offers risks, discoveries, and new and unfamiliar lands and experiences that accompany these journeys. None of these is as challenging as deeply getting to know and *truly* being able to love and accept ourselves with patience and compassion. One central point to my thinking is that we need to get things right within ourselves before we will be able to truly tackle these outer levels of social change, never mind undertaking exploration. As the adage goes, before we can really love another person we need to be able to have this core positive relationship with ourselves. This, in its essence, describes the work I have championed in the discovery and exploration of the unconscious. Much of my work has been focused on understanding and seeing what is required to deeply heal each of these layers to stably and sustainably evolve to larger, healthier, and more coherent compassionate systems. This landscape of conceptual transformations brings all of these challenges and more to light. If I were to summarise my thesis, I would describe it thus: To healthily journey outwards, we need to sufficiently build inwards, to understand and correct the interlacing of the triggers that create chains of distress and dysregulation. Understanding how to heal these factors is a critical point of change that can profoundly and positively transform the health of these inner and outer ecosystems.

However, part of the challenge is that this also requires simultaneously voyaging outwards as well working on both of these problems at once. In this writing, I would like to make the case for a renewed appreciation and similarity for the journey outward for the discovery

***Figure 6.1* NASA photo of Apollo lunar landing**

Source: www.nasa.gov/sites/default/files/images/574954main_GPN-2000-001140_full.jpg

of completely new lands that is incumbent on us actually understanding our inner emotional lives and how we do not need to be unnecessarily destructive. Moreover, I would like to create an appreciation for being able to visit our internal worlds and experiences of our relationships in a way that may be just as revolutionary or unexplored, wherein the interconnection between developing new levels of health and understanding will allow us to be able to make these quantum leaps within our technology as well as our perspective taking. Both of these require courage to truly get to know and understand what needs to be changed and healed. From my understanding, this would be a completely new event and unchartered within human history.

In thinking more about internal exploration, it is impossible for us as fully sentient beings to understand ourselves, if we have no thoughtful relationship to anything that is bigger or beyond us. I hope to clearly articulate some of the difficulties we have as individuals as well as a social collective in grappling with these experiences that eclipse us in many different ways.

The outer world

There are many aspects of modern life that Freud would have no knowledge about, from the landing of the Apollo moon missions, modern communications, computers, to the globalised feeling of interconnectivity to what we take for granted now in 2017. However, all of these historical facts did have their origins in the conceptual foundations largely laid during his lifetime, and these have matured in their technological evolution to where we very much are at a turning point in interesting and challenging ways. For example, we have artificial intelligence (AI) now that has the capability to drive a car completely autonomously or to think about itself. In 2015, in what is considered to be the most complicated game presently known, Go, the computer 'Deep mind' beat the world's best Go player 4 games to 1, when it had been considered unthinkable that a computer could beat the most expert human player. Part of this book aims to look at these complex and precarious points of change we are currently undertaking. Specifically, there is a difference between having AI, which we are able to control, and a 'thinking' machine, which might not want to be 'unplugged' or controlled by human keepers. This is a transition point that does feed into the health and intentionality of what we are creating and how this is likely to impact what is probably going to happen with these change points.

Figure 6.2 **NASA photo of 3-D printer**

Source: www.nasa.gov/content/international-space-station-s-3-d-printer

Being able to compare and contrast the different historical challenges within various epochs of time highlights the unique sets of questions we have to understand. Freud may not have had his own version of our UK astronaut, Tim Peake, who lived for eight months on the International Space Station (ISS), but he did have trends during his time period (Freud lived from 1856 to 1939) that lit the European imagination to some of the possibilities of exploration beyond Earth's surface. Jules Verne, the French novelist and the grandfather of the science fiction genre, helped open some of the first thoughts about what it might actually be like to embark on a mission to another place. Verne wrote many books on adventure exploration, including *From Earth to the Moon* (1865) and *20,000 Leagues Under the Sea* (1870). These beginning literary works inspired later thinkers to write about space travel, like H.G. Wells and *The First Men in the Moon* (1901). During the latter part of this epoch, the Antarctic explorer, Ernest Shackelton (1874–1922), set off for a series of three Polar explorations. Roald Amundsen (1872–1935) was the Norwegian Polar explorer who made the first successful expedition to the South Pole (1911). His work, along with Robert Falcon Scott and Shackelton, encapsulated the heroic age of Antarctic exploration. This was the closest modern equivalent to the risks, extremes, and expertise required for an expedition mission that approximates what current space mission training and requirements would be. A mission to go back to the moon or to venture to Mars would redevelop this notion of our modern version of this kind of new heroic age of exploration.

Competition to be the first or the best has been a consistent theme throughout exploration. The first artistic works of adventure and the Polar explorations inspired another generation of original scientists to actually develop this completely new field of application. Professor Wernher von Braun (1912–1977) was the rocket and aerospace engineer of the V2 rocket, which wreaked havoc across Europe in the hands of the Nazis. Von Braun became the father of modern rocketry design and laid the foundations for most of the same principles of the launch technology that is still used today. Von Braun had his murky career history working for the German Nazi war effort expunged after WW2 and he was reappropriated by the Americans to establish the North American Space Agency (or NASA). He later became one of the most seminal thinkers and inspiration for the biggest projects that NASA has undertaken. More specifically, he was in large part responsible for developing the Apollo space mission (1961–1975; with the first moon landing in 1969), and even developed the concept of space camp to train the next generation of astronauts that continues today in Huntsville, Alabama. Von Braun's deepest dream was to be able to help pioneer a space mission to Mars. The social-political tensions and competitive spirit between the US and the then Soviet Union spurred on the more recent version of the space race. The launch of USSR's Sputnik, and the successful space mission of the cosmonaut Yuri Gagarin, opened up an all-out superpower competition to develop supremacy over outer space. This pushed for the allocation of resources to make these significant strides to go into space.

These imaginative ingredients were all present in their earliest and unformed creative concepts when Freud was alive and working. However, it is similar to our own transitional period of space development, where relatively we have not explored even another planet, never mind billions upon billions of other planets, stars, and other celestial bodies. These more recent historical events for shorter jaunts to the moon, the International Space Station (ISS), critical Telescope Astronomy, and satellite gathered exploration have catalogued the next critical stages to provide the stepping-stones of greater exploration to emerge. The Apollo moon landings, the NASA space shuttle launches, the space imagery from the Hubble telescope, or the Mars rover landings need to be contextualised within a much larger footprint of the potential of a multi planet sustainable social system. None of these larger innovations had yet taken place when Freud was alive. Even within our own lifetime there has not been a very clear and consistent approach to the goals for space missions. In talking with my NASA and ESA colleagues, they readily admit that political uncertainties and changing priorities make bigger, more complex projects nearly impossible, and therefore (up to this point) have precluded any more meaningful larger space missions. Much of the progress that has been made has been done in fits and starts with no clear movement forward. This is why I feel the fears and uncertainties about our own natures, as well as what we might encounter in the outer universe, have slowed down, or even stalled, the creation of a reliable, stable, and predictable collective space programme that could grapple with these problems.

Context

It is not entirely correct to say the only relationship that the ordinary person has to interplanetary space exploration is to look up into the night sky or to sleep on a Tempur mattress (that was one research outcome from NASA). Our use and reliance on technology is integrally connected to these principles. However, it is true that few of us mortals have yet to go into low Earth orbit, and besides the lunar landings, no one has yet physically been to another planet other than Earth. One of the reasons I have wanted to write this book is to help link up a clearer understanding of why these principles of space exploration are such a key part of our collective inheritance. This is an interconnected problem, where we need to address certain core underlying problems within our individual as well as collective social psyches to successfully technologically innovate. However, we need to realistically work on these hard problems of how do we get to another planet like Mars to gain this kind of increased perspective. We also have to do this without having the mission crew going mad or inducing mass panic within the larger world population. Furthermore, much of this work centres upon the challenges of working on the subtle issue that we need to understand more specifically why we need to own a greater connection to the specifics of evolving from one definite place within the wider universe.

Salutogenisis

Definition of Salutogenisis. . . . Salutogenisis is a stress resource orientated concept, which focuses on making improvements to health and gives one the answer of why one's health remains well despite stressful events.

One of the big challenges is to look and notice how much of a relationship (or lack thereof) we actually have to the larger celestial world. It is there, but like many aspects of life, these meaningful experiences can be neglected and even missed in their entirety. There is a wonderful vignette in Don Dellilo's novel, 'White Noise' where the son, who is sitting with the father in the car when it is raining, denies that the rain drops are present as he believes the radio weather forecast over his own actual first-hand experience.

This need to pay attention to, and even more to develop a meaningful relationship, with the outside natural environment is actually part of our make up and who we are. If we can begin to properly contextualise where we are from, then it is easier to feel we have a better understanding of who we actually are. I would like to try to begin by giving a kind of tour to open up our appreciation of some of these aspects of our experiences that we can begin to appreciate within our own quadrants of outer space. If we first focus on a virtual tour of some of the major landmarks of our own neighbourhood of space, this can begin to build some of the context for us to think more about these issues of who we are as a human species.

As I began my writing, the moon is one of the easiest objects to see in our night sky. Some astronomers have said the reason that there is any life on our planet at all is because of the moon's orbiting the Earth. In doing so, it creates a tidal alteration of the ocean by dynamically moving and energetically gravitationally agitating the Earth's oceans, giving life to the cycles of water, weather, and convection currents. Many processes of growing are essentially linked to this process, as water is one of the crucial ingredients to life on our planet. For instance, giving water cycles, and the orbital shifts providing this movement, is equivalent to providing a giant ladle to mix everything up, giving energy and preventing death-ending stasis.

Wherever we go in the world the moon's presence proceeds us, with its waxing and waning, its solar and lunar eclipses, and its changes of colour from white to red, or even what appears to be a faint pale lapis tint. It is easy to imagine how pre-civilisation was able to discern time, where days and years could be organised through this reliable and consistent pattern of celestial presence. These dynamic qualities, particularly for females in their monthly fertility cycle, are very closely linked to this same 28-day lunar progression. It is interesting to think that these aspects of the environment within which we have evolved have these effects on our bodies, our sleep/wake cycles, and so many aspects of our physical builds with gravity, and even with our specific tolerance to Earth's temperatures. Our DNA and our genes have actually completely developed from being in a specific place. Equally, plants, especially flowering plants, evolved roughly 350 million years before us human beings (compared with our being here for roughly 2 million years). So every aspect of our development needs to be situated within this very complex web of evolutionary development within these parameters of Earth's astronomy. The interrelationship of the details of how all of these electro-chemical relationships have formed and are sustained really is something extraordinary. I am constantly in awe that we are able to walk outside without burning or freezing. Even the possibility of undermining our ozone, or climate, and the prospect of possibly losing some of the diversity and complexity that has taken hundreds of millions, if not billions of years to create, fills me with

Figure 6.3 **NASA photo of a particularly beautiful moon**
Source: www.nasa.gov/sites/default/files/thumbnails/image/37912591295_7c30a77efd_o.jpg

dread. It is truly a shocking thing to contemplate what the actual effects would be without an atmosphere.

The nature of energy

The study of physics is all about the use and conservation of energy, while astrophysics is the study of physics of the stars and the larger cosmos. Energy abounds all around us, and there are many levels on which this kind of energy exists from the minutely small quantum level all the way up to how energy works within an entire universe. The understanding of energy requires tremendous work to join up to see the relationship with any clarity. This quest by professional theoretical physicists still remains a mystery as to how the scales of the subatomic relate with the scales of the large. Sometimes, I feel that we do not really think much about how we are the creatures that have evolved within these multiple scales and how much we require energy to keep going. Thus, we do not resonate with a very unique and shared celestial space that is all around. There are many key heavenly elements that connect us to a better appreciation of the challenges inherent in living on the thin edge connected to the wider outer universe. The aura borealis is one of the most visually fabulous phenomena. My fascination with the moving phantom ephemera of the glowing electric neon green ghost of the northern lights is one that is both emotionally and intellectually compelling. This phenomenon is caused through the interaction of the Earth's magnetosphere being relentlessly buffeted by the space and cosmic radiation of the sun's rays, plasma ejections, and radiating energy smashing together into the Earth's magnetic protective shield connected with our atmosphere to create the greatest light show on the planet. Because of the Earth's molten metal core, it functions like a gigantic magnet with polarised ends and any energy coming in will be defrayed with this energetic process. These are the bioelectric fields, and their resultant effects are of a living world where this energetic protective system shields and cloaks our planet in something that keeps it (and us) relatively safe from the radiation and cosmic rays. When they are combined, they create a surreal haunting light show that can reshape our most memorable night of stargazing.

A key ingredient and instigator creating the aurora borealis is our own sun, which is 92.956 million miles away from Earth (Taylor Redd, 2012). Our own solar system's sun is comprised of a liquid hydrogen molten surface. Sunspots are magnetic field flux regions of cooler surface temperatures compared to other areas. When these regions become unstable, the sun is capable of solar eruptions where solar coronal mass ejections (CME's) create varying colossal explosions of orange-red liquid plasma that shoot out from the sun's surface (Aschwanden, 2005) like the largest volcano without the earthen crust shell to contain it. It is understood that over the full course of its life, our sun will expand in several billion years and progressively swallow its closest planets, eventually including Earth. She will be engulfed with a furnace of nuclear heat where the sun's expansion will continue into a larger and redder swelled solar death. This expansion will occur until the energy finally swells so much that the sun will implode in on itself,

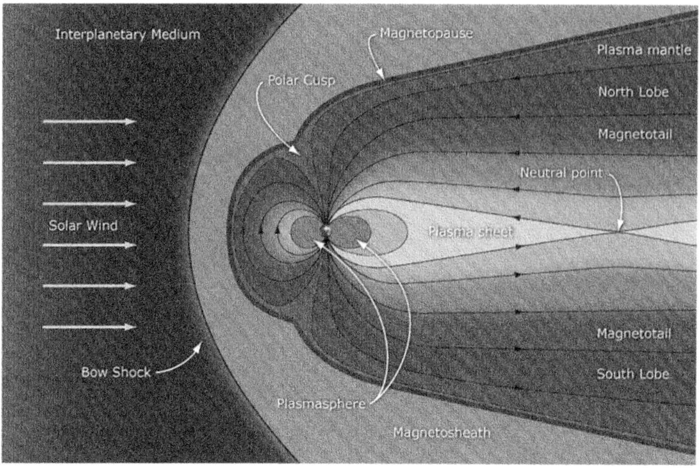

Figure 6.4 **NASA diagram of Earth's magnetic interaction with the sun**
Source: www.nasa.gov/sites/default/files/images/517890main_Earth-Magneto sphere.jpg

creating a white dwarf star. It is humbling for me to think that every person, like every planet, and most likely the entirety of the universe has a birth as well as a death. Each of these time spans and the passage of each life are carried with it, on its own course, as well as on its own developmental pattern of unfolding and perhaps its own deeper identity. These deeper structures of these apparently diverse lifespans are probably more similar than we may at first appreciate.

As we move out into deeper space, the combination of the unimaginable energy of destructive forces and the sublime beauty of nature both increase. This challenging notion of what is near versus what is far in space can play with your perspective. The inky blackness of the night sky is punctuated with the sharp contrast of varying glittering intensities of pure light that can accumulate to form the star belt of the Milky Way. One night, when I was camping with no light pollution, it became evidently clear that this seemingly never-ending star filled heaven surrounds us. I remember the comets whizzing by, combining the beauty on Earth and the galaxy with such an unimaginably limitless vista of places to explore. We modern humans miss much of this connection with deeper space, as we are very disconnected due to light pollution and a separation from the outside world.

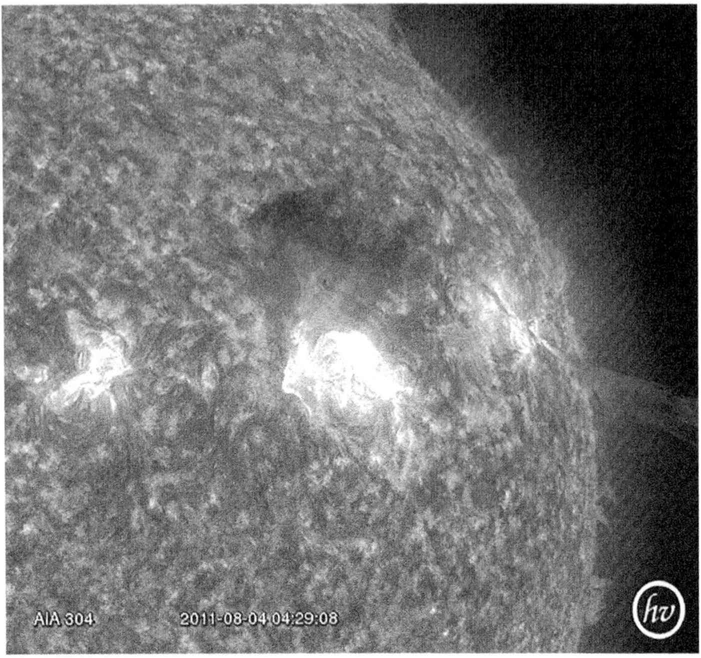

Figure 6.5 **NASA photo of close up of solar flares and solar surface**
Source: www.nasa.gov/images/content/577573main_2011_08_04_04_29_00_AIA_304-670.jpg

It is surprising to think that each of these dots of light is a possible planet or sun. With the time it takes light to travel, it is possible that these are the equivalent of celestial archaeology. For instance, there is a distinct possibility that these planetary bodies or suns could be extinct, as the distances are far enough away that even if we are able to see the dot of light from the emitting luminous source, it could have been destroyed many millions of years ago. However, we are none the wiser, as this information has not been updated to our location in our own light field. These extremes of time and spatial distances are amazing to think about. The wider cosmos has predated human consciousness and forms part of the primordial make up of who we are in relationship to the greater vastness of the collective of everything. However we learn, this scale, and our understanding of it, requires critical and constant revision. Partially, as we have come to learn that space and time itself can expand (and potentially can contract), such that the completeness and insight of how our understanding of these conceptual models work, as well as their possible implications, can change the explanatory level we are capable of reaching. These conceptual innovations require updating our science and technology, which in turn can create quantum jumps in knowledge (as well as within technology), thus helping us to refine and rewrite the laws of how these fundamental forces can ultimately be construed.

There is a description that everything is stardust. Or more precisely, gas, dust, and ice. This matter can also be described as inorganic – Calcium-Aluminium-rich Inclusions (CAI's, which are formed from solar plasma ejections a little like volcanic rock that forms meteorites, but these rocks are formed from suns) – and organic – for example, Polycyclic Aromatic Hydrocarbons (PAH's, which are formed when organic substances are exposed to extreme pressures and temperatures) (Vita-Finzi, 2016). It is likely that the similar concept of forensics applies with these phenomena. The principle of what is left behind in the interaction always leaves a trace and the transfer. Especially the interaction between organic materials carries with it the possibility of seeding the life-giving properties in another place (given enough time and optimal conditions). The transfer of these organic materials works something akin to interstellar bumblebees transferring organic life into other environments to seed the smallest potential of life within these environments. These are some of the key building blocks for the solar system, which extend the possibility of organic life finding other places to develop. The constituent raw materials and the assemblage with extreme star furnaces and absolute zero temperatures have an ecological cycle (and recycling) process that is similar to what we are used to thinking about in our own water and Krebs cycle here on Earth. These processes of building up and breaking down are taking place all of the time. The stability and ecological optimisation of life depend on all of the components that are needed to happen on so many layers and scales from the nano to the galactic.

The more that I have learned about these constituent factors, the more it has made it impossible for me to go outside, breathe air, and go for a walk without feeling in awe of how much is needed for all of these factors to work so well. This degree of living essence is something that each of us should neither take for granted, nor forget to hold on to with some appreciation. I was moved when I was researching more into this area and one Indian astrophysicist, Beg (2009), described the essential function of black holes, which like an interstellar root system breaks down space material. To provide some background, a black hole is described by the dictionary as a region of space having a gravitational field so intense that no matter or radiation can escape. However, according to Beg (2009) there may be significantly deeper processes going on with this process.

The nature of reality

There is some debate as to whether low-level Hawking radiation may leak or not. However, these gravitational siphons create energy transfers that are long lived, but will eventually disappear. The complexity of these smaller subsets combine together to form a large black hole, with the energy involved to create a gravitational field strength so intense that within this model space-time is understood to fold in on itself and produce what Susskind and Lindsay (2005) describe as a holographic universe. Developing from Maldecena's model (1998) that the universe is made up of exceedingly thin vibrating strings of nine dimensions plus time, where the majority of what can be interpreted as space is infinitesimally thin, the physical world, according to this theory, does not cover the complex vast space we think it to be. This concept has been taken further to hypothesise that all space is merely a simulation where we interpolate the volumetric dimensions of space (Bostrom, 2003). I personally do not feel this is the case, but it does bring into greater accessibility the relationship between possible internal and larger external spaces where some physicists have taken this version of our understanding of space, time, and ultimately of reality and stated it could be completely generated or even synthesised.

I come from the perspective in understanding the deeper nature and the interrelationship of space, and our understanding of these processes necessitates we need to deepen our appreciation of how these forces work together. It is challenging to wrap our heads around the scale of what we are looking at as well as looking at how interactive the cosmic relationship to sustainability and balance of the universe becomes. The scale of this 'cosmic' ecosystem dwarfs the singular planet model of ecosystems that we have been taught to focus on. This is in no way to say that we should avoid taking care of our own Earth and practice all of the aspects of good ecological stewardship. Far from it. However, I have not really seen a bigger view of these interdependent

Figure 6.6 **Milky Way with Airglow Australis**

Source: Yuri Beletsky (Carnegie Las Campanas Observatory, TWAN) https://apod.nasa.gov/apod/ap170203.html

processes of renewability. It is only in beginning to properly begin to grapple with these cosmological questions where my appreciation of Earth's environment seemed applicable to the possibility that there may be equally as complex an ecosystem within the universe. We just have not been taught to look at this level of interactive process.

One exception to this greater cosmological ecological viewpoint that I have come across is Beg (2009). He described his understanding of black holes from another perspective as a physicist, explaining that black holes work like the cosmological equivalent of plant and tree roots breaking down matter into more basic elemental material that can be recycled and put into the creation of new material to make up the greater universe. These astral processes of regenerating are similar to lysosomal cells in the human body. Both break down material; in particular, the black holes disintegrate dying planets of lifeless stars into material that can be transformed into the ingredients to form the birth of new planetary bodies. The universal level of cleaning and sustainably taking out the detritus as well as putting back into place a rejuvenated version connects together larger cycles of complex regeneration. I was very moved to think of this degree of integration of renewal as part of a universal scale of recycling that is likely to be at work. Perhaps these processes are operating at every level of resetting the disequilibrium that occurs with use and age. Annie Jump Cannon and Edward Pickering developed the contemporary Stellar Classification, The Harvard Classification Scheme, to describe different types of stars and to sort out the spectral lines (i.e., which chemical from the periodic table would be used depending on what spectral lines were showing). This work was done in 1901 and is still used today with only minor modifications.

I find Beg's work so interesting because it highlights the possibility that the same processes exist on so many levels in how the dynamic reintegration of materials, as well as energy, happens across many domains. In this version of stellar recycling, much like the reparative function of dreams in how we recycle our memories and experiences and feelings and the cleaning process of plants and the water cycle, even black holes have a similarly scalable regenerative process that takes these constituent parts and finds a way of reprocessing and breaking them down into parts and processes that can be used again. For instance, consider where the balance of sustainable renewal can be respected and not co-opted in a way that undermines the deeper health of that stable system, and the equilibrium of the integration of these deeper systems of life match up where one part is not tipping the sustainability out of kilter. In this way, all levels of life and what elements that life is made up of, build upon the need to be in a healthy balance with what is going out, what is coming in, and what is needed to be replenished. Nature is very good at naturally finding an optimum relationship between these competing demands to keep these factors in balance. One of the biggest problems we are encountering, especially within this time period of the Anthrapocine (The Time of Man), is that we do not heed what needs to be in balance and how to ensure these limits are respected.

From my understanding, even some of the models of how reality is described appear to be dysfunctionally depicted. For example, these processes, rather than the view of a hyper-flattened, false, or artificially created notion of reality (specifically, there is *no* actual space, time, or reality – it is all just a created artificial hologram), is the perspective that I think lends itself to the possibilities of dismissing or even abusing others in a dissociative and potentially cruel manner. The sense of a much deeper integration of a natural organic system requiring renewal, and recycling of all components of matter, energy, and life intuitively makes sense to me and provides the possibility of a sense of meaning and connection of life and aliveness.

This much larger interlocking system of natural ecosystems likely work literally on every level with this kind of dynamic and natural relationship, even on the planetary and cosmological level. Having a natural notion of the larger ecosystem would help facilitate a more settled space to think about and even explore these new territories. However, I think one of the reasons that our technological challenges have been held up is that we are working from a pathological understanding of systems or processes. These unhelpful views would quite understandably slow or even stop the innovation and conceptual jumps necessary to evolve a working system that could take us to the next level. To navigate some of these challenges, from my understanding we would need to put together a manageable and clear plan together to harmonise, and from my perspective we would need to create a two part plan that equally requires the other part of this process to be more clearly in place for us to be able to make these changes. First, we need to have a clearer and healthier notion of who we are as a species and also what our shared goals are (as well as to work towards these aspirational goals). It is critical that these goals be more shared, significantly healthier, and more compassionate in their nature. Second, we need a clear goal to move toward a coherent and stepped exploration of space where we need to pioneer clear steps to work on directly addressing these challenges to make this outward journey possible. The greatest challenge besides the physical effects of the medical and psychological impacts on astronauts is the incredible distances involved when travelling to more inhabitable planets.

Emptiness versus life within space-bacterium

As I write this, thus far, we have no evidence whatsoever of any biological life existing definitively elsewhere. There was one world-sensationalised scientific mistake where a meteorite collected from Antarctica was thought to have contained a possible fossilised microbe or bacterium from Mars (2009). This was later found to be highly unlikely and probably due to a micro bubble or other anomaly in the surface preparation of the rock (as it needs to be sprayed with a thin covering of liquid metal so it can be seen by a tunnelling scanning electron microscope). This experience was the first potential scientific proof that may have more clearly indicated scientific biological life on another planet. This could be contrasted with Orson Wells' War of the Worlds, where his radio programme aired a spoof on Halloween eve, October 30th, 1938. Wells aired a radio programme about a fictitious attack from outer space aliens laying waste to New York City. There is some debate as to what degree of panic or hysteria actually occurred, but the radio programme did create a sensation, where some listeners did think that New York was actually under strike. Much of the trauma literature of war and natural disasters (like the prolonged disruption of floods) establishes that if one's home environment is disrupted, that this can severely disrupt a sense of safely and coherence. This is the fastest path to creating defacto traumatic responses within a population. These issues of the psychology of one's environment can have a significant determinant effect on one's individual or social health and functioning. This is why the psychological challenges of responsibly going to another place (especially with such a completely revolutionary change as establishing a home on another planet) provide completely new challenges to the models of human and environmental relationships and concepts of well-being. Consider important questions like, what might happen if we get there, especially if there are other cognisant inhabitants? These questions raise potentially incredible deep anxieties of mission failure if mistakes are made with these foundational and critical steps.

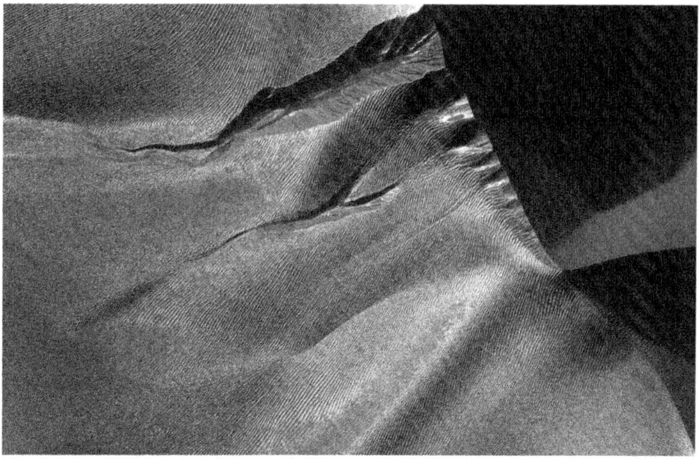

Figure 6.7 **Close up of surface of Mars: Gullies on Martian sand dunes, like these in Matara Crater, have been very active, with many flows in the last ten years**

Source: www.nasa.gov/sites/default/files/thumbnails/image/pia22349.jpg

The greatest challenge at the moment is overcoming immensely vast corridors of empty space, and as far as we know much of it is devoid of life. A brilliant quantitative calculation, the Drake equation hypothesis, was derived about the possibility of specifically estimating the probability or likelihood of life on other planets. The mathematics of examining the galaxy were based on the optimal conditions of being able to sustain life, and therefore, what were the real chances of finding life there? The first problem is thinking like an astrobiologist, rather than a regular cinemagoer. Separating out how the building blocks of life are organised, versus, looking at maturated and evolved processes or systems is required in looking at space science and deep space psychology. The definition of 'virus,' organic material, plant or otherwise, is very different than looking at a planetary weather system, or what technology that could power a spacecraft. Basic building blocks of a stable system of life to occur 'are required.' The Goldilocks distribution that is neither too hot nor too cold, but just right, must also be present; astronomers have worked out the optimisation for the patterning of the universe with the creation of stars. Astrobiology and planetary science have predicted that the 'Goldilocks band' of optimal conditions does appear to hold true where life should be predicted to exist as with our own planet Earth. However, because of the extraordinary vastness and distance to other star systems, we have no concrete proof of life (yet), other than our own. And here we are.

From what we currently understand about planet cosmology, it is self-evident that in the great majority of celestial and planetary bodies, life is also not an immediate given. I wish to bring these two aspects of optimal conditions for physical life and the conditions on an individual human level as well as the critical factors required for life on a planet to be sustained. These are both interconnected and mutually required. From my journey and research across the world, much of human suffering appears to be closely linked to this experience of a perceived absence of organic life and the potential suffering that comes from being deprived of life-giving sustenance and habitable conditions. From much of my clinical as well as humanitarian work, many of these truths can be boiled down to 'the healthier the environment, the significantly greater the potential for better mental and physical health.' In fact, this interlocking relationship of mutual casual interdependence really does require a healthy growing environment for nutrition, oxygen, and temperature regulation. If we are not eating vibrant and healthy fruits and vegetables, we are not going to have the necessary ingredients to be able to think or feel in a healthy way. If we put food in our bodies that is not healthy, it is likely to cause harm at every level, including undermining our thinking. It could have a detrimental effect on our ability to emotionally make sense of the world, or our ability relate to others will be compromised. If we are not getting the correct levels of oxygen, or if the temperature levels are not within a habitable bandwidth of what we are designed to withstand, this will make living and thriving a potential impossibility. Much of the cutting-edge research UCL did with their Extreme-Everest 2 research project, where volunteers' blood was taken on the 'Balcony' of Everest (7,100 m), found that for some of the participants had a blood oxygen level was functioning at 30% with all of the different factors such as oxygen de-saturation. The combination of life threateningly low O_2 levels, very low alveolar oxygen pressure, and extremely high CO_2 levels, combined to create environments that literally can easily kill. The combination of these issues highlighted the (short-term) survivability that was akin to critical intensive care unit patients that were previously seen as irrecoverable at this level of physiological functioning (NHS, 2009). This flexibility and core level of survivability sheds light on the exceptional adaptability of the body to evolve to extreme conditions. This is why the double challenge is for us to take the healthy version of what our bodies and minds require for us to grow and flourish, namely the right foods and medicines as well as finding suitable conditions, but we also need to address many of the negative thoughts and our prejudices that have caused so much conflict and distress so we neither unnecessarily contribute to nor create unhealthy living conditions, nor do we underestimate the core adaptability of what our bodies and systems can acclimatise to, given the right circumstances. This requires nominal balance and proportional thinking to get this kind of survival relationship right.

From my and many cosmologists' perspectives, if we do not take this next major step in exploration, we will be unable to ascertain this deeper perspective. Therefore, in missing this next step, it may be seen that we do not truly have the freedom, or maybe even the perspective, to be able to look at ourselves to build a relationship or a

Figure 6.8 **NASA photo of a single growing plant**

Source: www.google.co.uk/imgres?imgurl=www.nasa.gov/images/content/462887main_iss022e097231.jpg&imgrefurl=www.nasa.gov/mission_pages/station/research/10-074.html&h=2848&w=4288&tbnid=ktb-l97o_e9ORM:&tbnh=160&tbnw=241&usg=__WOau_HlFGfZwaiaXFC0ipEKZ0ts%3D&vet=12ahUKEwjlwY6AgorbAhVF6qQKHSu1AlsQ9QEwAHoECAEQMQ..i&docid=p3Jd49OwX2Z3RM&sa=X&ved=2ahUKEwjlwY6AgorbAhVF6qQKHSu1AlsQ9QEwAHoECAEQMQ

greater context of our true place both within ourselves as well as to the greater outer universe. Properly sitting with this deeper relationship does potentially transform many aspects of how we can improve our healthy dynamic relationship and stop creating needless suffering. Equally, many of the critical steps we need to integrate into our world-view for how we *must* take care of our planet (so we do not destroy the fragile ecosystem) requires this double layer of insight and stewardship. Looking at these questions from a deeply psychological vantage point, this impairment of working toward a healthier expressed capacity instead of routinely finding more pathological manifestations of societal expression will distort our own collective self-identity as well as undermining our capacity to journey outwards. This dual blind-spot and limitation will only increase if we keep making poor choices in our self-understanding and how we express a healthier capability can completely change what becomes manifest. Ultimately, this self-destructive mode undermines both our compassion and our fuller developmental capacity of expression of aliveness.

Connecting the interaction between our inner world and the outer universe: understanding the limits of our own world and its finite resiliency

We are on the brink of an entirely new era in human consciousness. As I have briefly mentioned before, we are now living in the Anthropocene epoch, the time period of human beings. How we think about key issues, how we work together, how we look after the planet, each other, and everything that is contained within this eco-sphere will determine most likely if we survive or not as a species. These extremes in managing stress, technology, and shared global experiences are all part of new ways to navigate one of the biggest steps human kind has ever taken. Going to a new planet is central among these critical steps that I understand that we will need to be able to negotiate and make sense of. Psychology and the way we understand how we think about our planet, our environment, and ourselves will transform the totality of integrating these different elements. We must find a way that is about working harmoniously together to find sustainable solutions rather than viewing these aspects as discrete, separate elements that are working in direct conflict with each other. Moreover, my recent work has been looking at the possibility that with potential resource limitations, it could be very likely that dysfunctional leadership, instead of trying to look after and better manage resources, will likely try to corner, dominate, and use up even more limited stores even faster to gain a greater competitive advantage over others. The precarious nature of these potential increasing series of tipping points really does require a more profound evolution to human thinking and collective capacity than we have ever had thus far in shared human history. Even within our geologic history undermining some of the basic principles of the water cycle, the tidal structures, and the protective aspects like our atmosphere will change with global warming.

This is important from a psychological defence mechanism. Consider the Nietzschean adage – when you stare into the Abyss, the Abyss stares back into you. There is no abyss more boundless than outer space. To my mind, this presents some of the significant challenges of nothing exceeding this level of the absolute. The human imagination begins to foreclose or falter rather than struggle to focus and succeed in working out the problem of finding reasonable answers to tackling and solving these crises (however many in number these disasters may be). Much of this stems from a crisis of confidence and belief in ourselves, our values, and the uncertainty of portraying these. Psychoanalytic thinking does hold the belief of the destructive tendencies man is saturated with on every level, from his unconscious to his conscious mental life. It is likely that this uncertainty and understandable fear about our own damaging nature has to contend with two possible outcomes, neither of which sits comfortably when tested.

The first dyadic conundrum: are we alone in the universe? This is unlikely to be absolute, but we may be so far away from other life that it is the practical equivalent to this first point. If the answer to that first question is that we are not alone, there is a second part to this problem. Much like the child who insubordinately wishes to challenge his father, the technological asymmetry of confronting another group that could potentially surpass us, contain us, and humiliate us, or worse, also activates our fiercest apprehension. Stephen Hawking, Professor of Physics, has publicly said that one of his greatest fears is of other life forms knowing of the existence of life on Earth and coming to exploit our resources and potentially destroying everything. I do not wish to create any dissent or possible cliffhangers, but my next book will try to grapple with these questions. The second factor, which I feel is much more subtle, does contain the kernel of insight into whether we are able to see and understand our individual as well as our collective species behaviour. Namely, our bellicose tendencies may be a gross overreaction and much of our behaviour causes ourselves, as well as others, tremendous unnecessary suffering. If we truly have to see, or more challengingly understand this behaviour from another's perspective, this would likely require a vast rewriting and shift of the predominately conflict entangled response that all of known human history is replete with.

Our intra-psychic life contains a quantity and variation of comforting as well as anxiety provoking experiences. Our environments will also similarly exacerbate some of our internal senses of identity. Space, with the absence of the usual environmental stimuli humans are used to, and with the significant risks and pressures, creates a unique set of triggers for the individual as well as for the social group. From the cases of what we have seen in Mars simulations, being so close to the edge of death can drastically drive this kind of fear as well as real scarcity to the fore. It is my belief that if humans are to tackle this next quantum step in exploration, it will impact upon our identity as a species as well as reach the limits of overcoming our double fears – those of the outer known world and perhaps the deeper one of ameliorating our ambivalence about ourselves and our deeper fears about too little or too much of what is in our environment, as well as what is the *true* nature of difference (of otherness)? In combination, these two interfolding layers of fear of ourselves as well as the fear of others drape over our shared social interactive experience, shrouding or potentially further dysregulating a vision of what we can work toward together. These polar opposites in our human nature of our attaching and competitive tendencies can trigger deep levels of potential violence that may be wreaked or perpetrated as opposed to looking at deeper notions of truly collaborative, shared responsibility and compassion.

7

Case example, mission to Mars and mission critical psychoanalytic processes to be realised

Practically mapping out the steps to successfully plan a mission to move toward travelling in and eventually colonising different parts of outer space is the holy grail of space travel. There are very different perspectives on how we can manage to break these tasks down into controllable and achievable steps to actually accomplish this vision. One of the most interesting thinkers in the field of space exploration and space psychology is Erik Seedhouse. He is an extreme athlete, an academic psychologist, and an author who has focused most of his research and work on all different aspects of space exploration, including writing several books on the theme linking historical aspects of exploration and its application for innovating space travel. His viewpoint is definitive that before anything like a planetary voyage is undertaken, we need to revisit and refine the work with the space travel to the moon that the Apollo programme had begun. This earlier work requires updating the technology and the work to establish a working lunar base on the moon. His thinking is that much of this is larger staging to lead to establishing a colony on Mars and requires understandable meticulous forward planning where many of the multiple layers use staging posts, and where infrastructure and supplies need to be gradually brought forward to make this a reality. To my understanding to extend the capability of exploring other planets requires humans to work to the complete extension and upper-most capability to achieve a new level of capacity. Moreover, planetary exploration will also need to manage to sustain these kinds of complicated functions for social survival for extended periods of time. This makes sense to me, where it is easier to trial these aspects on the moon, and where it is much easier to address problems at this extended distance, and moreover to work to try to perfect this level of inhabited continual space mission, than to work toward a singular one-off mission to Mars. There is a logical order and possible perfectibility to how one can really make innovations within the space realm. For example, by developing grapheme nano cables it would be possible to anchor a continual use space elevator that can lift cargo and for much larger space vehicles to be assembled in near Earth orbit. This would almost singularly bring down the prohibitive cost of space flight as well as reduce the extraordinary polluting capacity of burning massive amounts of rocket fuel.

The other aspect of his reasoned argument is that we need to have a very clear working model that is closer to the Earth in order to have the resources to ensure help is at hand if a mission gets into trouble. To my understanding of things, this sounds like a highly sustainable and considerably longer-term perspective to be able to ensure that each stage of the smaller steps is done responsibly and fits together, so we do not put the astronauts who are undertaking these arduous explorations in undue danger and have support structures reliably in place. By not trying to do something where there is a previously logical precedent preceding it of what fits together and what causally needs to follow it, it places perhaps undue pressure and expectation on those involved more than required. For example, aiming for a manned mission to Mars catapults world attention of anyone going up in a way that makes it more of a unique, almost bizarre celebrity status that does not seem to have a digestible way of having anything that compares. However, building a constantly inhabitable lunar base that acts like the next stepping-stone does make it easier to responsibly work toward reliable (and responsible) space exploration.

My own perspective is that whomever goes up for the first time, even 'if only' to the moon, will have so much scrutiny and intrusive undue interest into the person's or the team's life that this will function as another significant factor that will feel like everything is under a microscope. This will be even truer if we skip all of the earlier steps and go directly to interplanetary travel to have a manned mission to Mars. This more radical non-linear mission progression to go for distance, rather than safety raises extraordinarily complicated mission planning challenges and ethical conundrums. There is a real risk that if something goes wrong, the wider public opinion could create a backlash and potentially create considerable delays for a subsequent viable

Figure 7.1 **Astronaut space walk on the lunar surface Apollo program**
Source: www.nasa.gov/sites/default/files/thumbnails/image/a15pan11845-7.jpg

manned space programme. There is a general consensus with space medicine experts, space psychologists, and different space engaging experts that the physical problems of radiation, physiological changes like bone demineralisation, and longer-term health consequences pose a real and justifiable concern, especially at the durations of time and the extremes that a longer and more hazardous mission to Mars would indeed pose. A mission to Mars might take a few different forms. I have been involved with lectures and projects focused on one-way permanent settlement of Mars and also the two-way return round trip.

8

Space exploration to space travel, and the future of psychoanalytic insights into human consciousness

From spectator to astronaut

Really planning to go to a new planet

How does the planetary scale relate to us humans on an individual or social level? This is a tremendously complicated question, but returning to my inquiry about the nature of human space travel, much of this relates to the challenge we have to properly understand our relationship with the wider universe and thus to sustainably reconcile our nature within all of these challenges. The cosmological domain and the functionality of the human mind both require models for explanation of how each of these entities work. To grapple with these factors, many of these questions are just as philosophical and ethical as scientific in their nature. Here much of the previous corpus of my own psychoanalytic work in the features of the unconscious and the inner mechanisms of the human mind come into their own.

The relationship between the conscious and the unconscious probably has some of the dialectic in its quality of matter and anti-matter, or of energy and dark energy. The interaction of the external forces and their interface and impact do have an understandable relationship on the level of the individual, just as much as they do on the universal constant levels. The 'standard' cosmological model is in agreement with the scientific data of the recent acceleration phase of cosmic expansion (Uzan, 2010), which places a focus on understanding more complex models of gravity. It also adds newer additions to our understanding of the interactive fields of Lorentz's work on Strong versus Weak gravitational fields in how dark matter can logically alter the cosmological constant to create an expansion of these. That is, dark energy is the description of the astrophysics phenomenon of the universe's accelerated expansion in what appears to be a more dynamic interconnected integration of forces. It requires another shift of thinking. For instance, a more recent equivalent is the dawning imperative of the sustainability problems we are facing with the chronic abuse of the natural resources on Earth and how population pressures, pollution, as well as factors such as over use of fossil fuels and climate change have shifted in a similar way. All of these, to some degree, require a better understanding of problems of acceleration and how can we make sense of them so they do not feel insurmountable.

The point I am making in a highly condensed way is that the explanatory models we have to construct the realities, for the physics or even the psychological occurrences, need not only be written, but also require us to correctly apply these and to understand the nature of the present day reality (or even the future trends) and how we can create healthier versions of these aspects. I make this link from a psychoanalytic vantage point, just as our physics has changed with the awareness of cosmological expansion. There is no reason that there cannot be an equal opening of consciousness working to reconcile these discrepancies in the health and longevity versus all short-term resource use. This is why the explorer's psyche in looking at these difficult problems does inexorably link up with thoughts about space exploration. To put it simply, I see these as one in the same types of questions, where we need to address *both* aspects of these issues to find a stable and workable solution. One problem I can see, if we do not resolve some of these core problems with concepts and approaches, is that any intervention we try will potentially come out as dysfunctional. Therefore, it is important that we do not reproduce these kinds of problems in our voyaging and possible settlement of any new society on another planet, just as much as it is important for these core levels of dysfunction, injustice, and pathways for ill health are resolved here on Earth.

9

Radical transformations in concept and development of technology

Space as applied measure of human maturation

Space medicine, and the potential shift in actually opening shorter duration space missions and eventually moving to establish true deep space exploration, is an important next step. What are the potential impacts on technology, consciousness, and the practical lives of everyone around the planet and on the global imagination in moving toward to making this an actualised reality? It is important that we focus on the psychological experience of space and learning points related to what space can help us understand about human psychology on all of these many levels.

Space travel and the extremes that will be encountered by astronauts potentially within the next 50 years include the challenges of actually living on Mars with mega-wind storms, extreme temperatures, and 40–100 million miles of distance from Earth to Mars. How will technology and humans face these new, harsher environments, and how will it change how we think about ourselves and (thus far) the relatively more benign stability of Earth's eco-sphere?

What I am arguing is that there are potentially two critical psychological processes that we need to reconcile more fully to make these shifts. To reach these new milestones, the individual psychologies of the men and women who will be going to these new places will have to acclimatise and adapt to these realities; additionally, the larger social collective psychology of how the rest of the world relates to these challenges and changes will also need to radically develop. Our perceptual process will not be able to see the possibilities before us

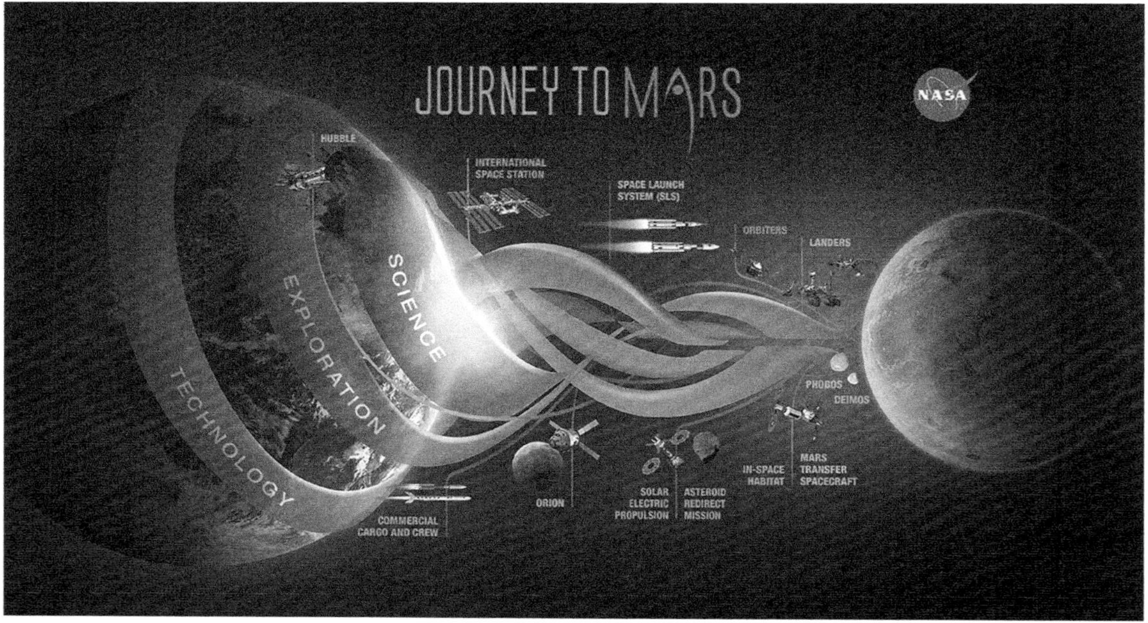

Figure 9.1 **NASA overview of manned mission to Mars**

Source: www.nasa.gov/sites/default/files/thumbnails/image/journey_to_mars.jpeg

until we can acclimatise and understand these new spaces. An example of this on more of an individual level is how the astronaut Chris Hadfield described his experience that he could not see the pen he was writing with that was literally right in front of his face. When he first got into 0-g environment, his mind had not wrapped itself around this possibility of a place without the effects of gravity. Therefore, he could not see what was right there directly in front of his own eyes when his pen was floating literally next to him. From my deep understanding of human individual and collective group psychology, I recognise that there is very likely a similarly shared process of emotional contamination and social thinking within the collective meme or shared social mutual perceptual understanding that is likely to be at work on this communal level. The collective mental blocks in thinking humans could go to another planet, never mind to live on one, is probably similar to the challenge of undertaking a marathon – where if it is believed to be impossible, it is likely to make it so. Now, the capacity of doing 24 marathons in a row through ultra-marathons has completely changed how human beings have understood what we can imagine and what we are psychically capable of accomplishing. Similar to any significant breakthrough or true revolution in thinking, something needs to profoundly be reorganised for these events and then new pathways to emerge.

In a similar way of crystallising how these deeper levels of problem solving can emerge, space offers us collectively very profound challenges in how we understand these shared social-emotional processes actually function. As well, these harder problems push us to see how we might find better solutions that can function more optimally as well as what are the blocking beliefs that we may not be conscious of that actually may be present within the systems at work. At its best, it is possible to achieve truly new realms of capacity with properly utilising disruptive innovation. For example, the interface with aviation and space psychology will bring in the psychology of human factors into possibly significantly improving flight safety and consistent flying capability within the even higher risk milieu of space flight. Aviation test assessment looks at how there is no greater challenge for human kind than the frontier of deep space travel, where learning from mistakes requires many more iterations of everyday flight than has already been logged within space launches. No single area of technology and innovation will push the physical, medicinal, and psychological limits of what we collectively understand regarding the possibilities of extreme environments. Deep space exploration, especially attempting to colonise a planet like Mars, holds one of the most unique challenges of extreme medicine. For instance, utilising the most technologically advanced kit while simultaneously working within the most hostile environments with potentially zero resources is a scenario that as yet we can only partially simulate. These scenarios of the edge of catastrophic failure and ethical nightmare scenarios require unprecedented problem solving. The same rigorur we take for granted in regular airplane travel does not yet exist where even with a successful launch and landing becomes only the first phase of many. The consistency where hydroponic crop failure could mean certain death rather than simply having to take chicken instead of having the beef option, if this ran out on my recent return flight from Japan on commercial air travel. The stresses for space exploration is an entirely different experience of potential deprivation and risk. What is the relationship between our everyday flying experience and space travel? How different would this really be and why? What practical and psychological frameworks and changes do we need to bring in to standardise space travel to relate to these very refined and established ways of working?

To answer some of these questions, we need to go back in time to the beginning and development of air travel. Like most areas of development, the advent of any new field carries with it much chaos, mistakes, and rich learning for how it can be done significantly better. This is the case with space flight and travel. From my understanding, the challenges are absolutely great for space exploration, however, the implications for correctly managing these issues are so much greater to push our human capacity forward. The challenges for what we have to grapple with regarding attachment separation and potential solitude to a level that profoundly triggers our human and social fears and anxieties. The standards and safety we absolutely take for granted within air travel require entirely new reengineering and a way to systematise these safety frameworks within space travel. In a similar way to aircraft test pilots with their hunger for speed and appetite for risk, the new breed of potential astronauts willing to go so much further afield to venture to Mars will require many challenging personality factors that need to be correctly understood. We can think about the personalities of those who have agreed and been shortlisted to go to Mars. If we think about the process of training and the preparation for a mission to Mars, it would be logical that this process would change the psychology of human beings. The incredible technical and scientific challenges that have never

Figure 9.2 **Astronaut carrying out research on the ISS**

Source: www.nasa.gov/beta/content/esa-astronaut-samantha-cristoforetti-2

before been explored directly connect with the ethical and human factor elements that include a necessity to tolerate real boredom, just as much as sitting with unbelievable and unrelenting chronic stress. This triggers our instant reaction, when hearing of these stories, to overlay their experiences with our own and decide, 'Could I do that?'

On one of the BBC radio interviews where I was one of the expert panellists, along with the astronaut Bonnie Dunbar as well as PhD student Hannah Earnshaw, who is shortlisted for the Mars 500, we discussed the one-way mission to Mars. It was interesting grappling with the issue of a one-way versus a round-trip mission to Mars. My NASA astronaut colleagues think that the one-way missions are ethically reprehensible, while proponents of a one-way mission, for example, Hannah, advocate the need for a more streamlined and realistic first step to make progress to get to another planet, even if it personally puts them at potentially grave risk for injury or perhaps certain death. I have nothing but admiration for what Hannah has discussed and what she is willing to sacrifice to make these plans a more realistic possibility. However, much of what may not be considered by many is the larger collective social zeitgeist of what getting a manned mission to Mars wrong might cause.

This is why I feel that the practice of bringing in and ideally listening to professional experts, like myself, to preventatively address some of the issues and to optimise success, is of the upmost importance. This will be one of the most focused and shared (never mind heightened) human experiences that we may have had in our shared history. Never before in documented human history have we worked to visit another planet. As phenomenon of shared human experience as well as having a voyeuristic sense of interest in novelty and exploration, this first experience at manned space travel will likely capture our collective attention. This will include the social media hype and the need to share, scrutinise, and analyse this meme of a novel human experience. It will be likely that for everyone involved, including those of us watching from Earth, as well as those going to the 'Red Planet,' the mission will ramp up the emotions like few events that we have seen.

10

Mutual evolution of concepts

New paradigms for the future of human development

A thought experiment from the beginning to the creation of a new era in human capacity is also fascinating. Imagination becomes reality for a successful space mission, in the challenges and creativity in taking these spaces from imagination and integrating our updated reality for these technological and practical changes of the interface between the true inner and outer relationship between them. We can relate how modern space exploration very much mirrors the possibility of our inner journeys to major transitions in shared human experience. These major changes to human shared belief with the Apollo space program foreshadow the even greater challenges and changes that will occur with the possibility of humans going to Mars.

We need to understand one of the most breath-taking adventures of what will happen in human history and how will this change our individual and collective consciousness and psychology. Within this epic narrative of a manned space flight, from the perspective of both a crewmember and a global initiative, we need to update much of our thinking. We need to focus on the emotional and cognitive pressures astronauts face, as well as the second and third order effects that they have to deal with. We will give a heart-stopping overview of the hard science and cutting-edge psychology behind the challenges faced by this new breadth of extreme astronauts and the communities they live in. The field of clinical psychology and psychotherapy offers new insights in looking at the impact of prospects like never seeing a tree, of disruptions in communications, the possibility of adapting to solar radiation, or the neuropsychology of low gravity environments and possible blindness or disability within high-risk environments, or the ethics and practicalities of severe limited resources. What do these mean for the courageous men and women who have signed up for this project – and what impact could the ensuing changes in perspective and technology mean for everyone else on planet Earth?

These challenges and understanding these aspects could potentially positively and compassionately improve sustainable life for the rest of humanity and help to address our increasing environmental crisis as well as redefine our view of ourselves. This lecture marks a new gateway to help us in interacting with the rest of the greater universe and understanding ourselves within it.

There appear to be two different divergent and potentially competing branches for innovating within space travel. The established governmental space agencies include NASA, the European Space Agency, the Russian Space Programme, and newer space contenders such as agencies from China and India. These governmental funded agencies have been pushing the boundaries to take existing rockets and professional astronaut selection programmes to problem solve all of these issues within the capability along with sustainability to reach

Figure 10.1 **NASA drawing of manned Mars modules**

Source: https://en.wikipedia.org/wiki/Human_mission_to_Mars#/media/File:Mars_design_reference_mission_3.jpg

an extended goal, such as making lunar missions or even the much bigger goal of focusing on going to Mars. One of the biggest problems in frankly discussing these goals with key members of the space community is the political challenge that makes finding consistent funding, which profoundly effects the clarity and reliable goal setting related to managing these vast projects, difficult. From my understanding, the experience of working toward these big space projects and trying to build toward a realised version in the face of having the lifeline of funding slashed, the focus of the research moved, or if the support public opinion lost can cause a rupture that can entirely cut the programme or change what is supposed to be accomplished. The frustration, if not feeling demoralised by this caprice, would be entirely understandable. This highlights the interdependence of finance, political anxiety of public opinion, and engineering problem solving. Each of these needs to be completely in line and supportive of the other, or as most government space professionals will attest, any consistent or visionary goal may start off to go to the moon, or to Mars, but by the time that a new government leadership is elected and rebalances the budget, that shared vision for the new rocket programme is replaced with a completely different and frequently more disheartening type of firing. It is important to note that there is a tremendous resiliency within most major government-backed space programmes, where this problem is becoming much clearer and the complete inability to run any effective space programme without having effective planning and funding (as well as having public backing) is becoming more apparent.

All of these issues outlined have increasingly opened up large private companies and entrepreneurs taking first steps to edge forward to build serious contenders and alternatives to the previously an all-government-held monopoly on space exploration and travel. Companies like Space-X, Blue-sky, Virgin Galactic, and many others are working to punch a hole in the hegemony previously held exclusively by space launches organised only by nation-state-actors. Many of the business giants that have created PayPal, Amazon, and Virgin Airways, as well as other large global business presences are bringing in commercial focus, and these highly innovative systems are making fundamental changes, such as pushing for highly aggressive product and technology development. These commercial endeavours are not only bringing in more consistent funding and freeing up much of the bureaucracy, but they are setting goals that are reminiscent of what appeared to be the impossible realisation of what President Kennedy, who threw down the gauntlet to have a manned lunar landing by 1969, wanted us to achieve. What Zubin, an outspoken advocate of the importance of a manned mission to Mars (and probable pioneer in developing the thinking of interplanetary space travel), has developed an economically mindful plan for a manned mission to Mars. Zubin has described, with the present bureaucratic approach that forms the standard governmental approach to space travel (as juxtaposed with launching satellites into near Earth orbit) is drastically selling short what human space exploration is actually about. The traditional space agency approach, because of the layers of political uncertainties and restrictions, has set onerous and omnipresent limitations that serve to stagnate these next logical steps in space innovation. Many have highlighted that even though there are fits and starts of progress towards these larger end states, it is surprising how problematic any smooth progress has been to build on the lunar landings and how endemic, with traditional approaches to this field, and how surprisingly limited the present accomplishments have been in delivering a sustainable level of space travel and exploration. From my own area of expertise, I would point to a kind of glass ceiling of human thinking to contemplate being possibly a multi-planetary species. Understandably, there would be challenges in resolving the conflicts that are already present, and we would be reticent to duplicate these on another planetary environment. However, it is more likely that if we are not stuck, or mired down in a potentially competitive turf war, instead of focused on a much larger shared collective undertaking, this would focus the shared intelligence toward a worthy and deeply innovative aim.

11

Connecting up the archaeology of the time, space, and psyche

During the histories of naval exploration, the possibility of discovering new worlds combined great risk with the possibility of fame and riches. The closely guarded state secrets of maps and the resources of seafaring countries would concentrate incredible national resources on these high-risk endeavours. Similar to ocean exploration, space exploration carries with it truly new worlds and great risk. For instance, consider the Drake equation hypothesis about the possibility of life on other planets. This equation predicts the probability of the likelihood of life in another part of the galaxy based on the optimal conditions of being able to sustain life. Much as I described before, with the Goldilocks distribution that is neither too hot nor too cold, but just right, astronomers have worked out the optimisation for the patterning of the universe with the creation of stars, anti-matter, which is a link to the unconscious and the deeper nature of energy.

For instance, the collective awareness of trepidation requires a quality of expertise in leadership and public communication to help everyone in the public domain understand the importance of becoming a positive stakeholder of a large-scale investment of public good will – as well as financial commitment – into a multi decade, multi-billion pound project: a manned mission to Mars. This would be the project that would be far enough to stretch us to the limit, but would still be close enough that it would be a feasible next step for an extra-orbital planetary investigation.

A question remains: 'How will space change our psychology?' From my understanding of really trying to deeply get to grips with this issue, I feel we are potentially on the brink of an entirely new era in human consciousness. I feel we might lurch forward, or at least potentially divide from the 'Anthropocine' Era, to develop into what I would name as the 'Galactacine' Era, or space-fairing humans. These extremes in managing stress, technology, and shared global experiences are all part of new ways that we need to navigate one of the biggest steps human kind has ever taken – like going to a new planet – which would then open up, in time, the rest of the larger environment of space. Psychology and the way we understand how we think about our planet, our environment, and ourselves are having seismic impacts on every aspect of our lives and the future of our species as well as the planet.

A little more than a year ago when I was on a specialist panel at the extreme medicine conference addressing the challenges of a two-way mission to Mars, we panelists were asked what positive benefit such a mission might have. Was it merely a vanity project? I responded that I understood, besides putting in place a back up plan for other options for habitable environments (which should not be underestimated), that the technological innovations that are required for such projects create disruptive innovation within spin-off technologies. These new ways to integrate and synthesise data and apply information to improving real-world problems form a significant puzzle piece that would not have emerged without these larger scale investments and multi-disciplinary endeavours. The challenges within these potential levels of disruptions from these technological innovations create, as well as solve, just as many problems. These require deep patience and responsible ownership to work through. For example, a lot of the technological communication and computing improvements have placed on our doorstep the reality of AI (Artificial Intelligence), and robotics are putting within several years the probability of realised automated driverless vehicles (for example, with lorry trucks) and replacing many factory jobs (as well as some skilled occupations) with AI-enabled robots or computing processes. You might ask what this has to do with space exploration. I would describe that much of the technology that has to help us go to other extreme environments like the moon, Mars, or even further planets requires innovations that the first iterations of these remote operating platforms will cause to come into existence. Many of the Voyager satellites, or Mars rovers, represent the first time these hugely complex computer programmes, remote operating systems, and autonomous extreme environmental technologies were created and directly applied to being used in these most challenging of circumstances such as space exploration. Every step of the engineering for these projects necessitates that each of these objects and everything that is required to go into them is engineered so they are built to withstand unbelievable heat, cold, severe cosmic radiation, wear and tear, as well as challenges of stresses and pressures that redefine the concept of resiliency.

Figure 11.1 **Mars rover on the surface of Mars**
Source: www.nasa.gov/sites/default/files/images/414002main_r_spirit_full.jpg

One of the factors of the design pathway process is that there is an underlying step-by-step evolution to how all of these aspects are methodically fitted together. For example, frequently, some of the science from theoretical physics or astronomy makes certain predictions, or there are findings from these disciples that require further work within allied disciplines such as astro-biology or astro-chemistry; then there are experiments that some of which can begin on Earth, but to truly drill down to clearly understand these principles, further work is needed to go to the next stage. These further levels of examination require more in-depth applications of space research instruments. For example, staring into the night sky needed to be improved with Earth-based telescopes; these eventually required improvements to be supplanted by extra-orbital radio telescopes like the Hubble. To reach the next steps, for instance to study the viability of mathematical approaches to space science, like Drake's equation to look at probabilities of habitable planets, requires that we are able to look with unrivalled levels of clarity into far distant solar systems with even newer and considerably bigger orbiting distant space telescopes to properly investigate more about the presence and viability of specific sustainable and habitable exo-planets. These kinds of quantum jumps in telescopic capability have begun being designed and constructed to make these new leaps in scientific measurements, planetary chemistry, and astro-biology eventually possible, to allow us to properly grapple with not just the fundamental building blocks of primitive life, but what is needed to ensure liveable ecosystems.

For this to actually work, some of the most mind-blowing complex science is required to bring everything that we have come to study integrated together, and moreover to absolutely extend the research to reach a new quantum leap in problem solving for each of these fields. Many of these improvements in joining all of these areas up requires continuous breakthroughs in the discipline of physics. Physics is the one science that brings together the most minute as well as how these processes may be working on the scale of the entire universe. At one time to hold together and simultaneously focus of the very large with examining other solar systems are being developed in parallel with disruptive innovations into the completely invisible quantum world. This kind of work is typified with bigger supercolliders, such as in CERN with its particle accelerator being built and the fundamental forces of matter tested. There appears to be an increasing understanding of a tighter and more integrated approach of how these various forms of physics and space science are integrated together. These improvements force connections between how the research community can understand the theoretical underpinnings of these aspects as well as how these can be applied to real-world applications in expanding both our understanding as well as making these explorations a viable possibility.

There is a repetitive process through these very demanding and challenging issues, to have to go back to re-examine these fundamental laws of physics, chemistry, and biology, to make sure these forces can still operate as predicted, without new constraints or exceptions falsifying the operating parameters of what we know. The changing scale or circumstance for the quantum level is still not clearly correlated with how these large-scale processes operate. This is where space exploration and cosmological research provides one of the best testing grounds to properly fathom out these core scientific truths. Each step forward into these ultra-extreme environments necessitates a testability and refutability for each layer of how these laws are understood to

Figure 11.2 **NASA photo of deep space and spiral galaxies**

Source: www.nasa.gov/sites/default/files/images/611265main_hubble_holiday wreath_full.jpg

run. Space exploration is therefore tightly linked (at least in the government space agencies) with being inseparable from research aims. It will be interesting to see how possible privatisation of space exploration will alter these tenants that have been part of the backbone of most of the history of modern space science and space exploration. In summary, for this point, there is a process that stems from the science that requires the innovations for technology for unmanned improvements and testing for these theories, and that eventually to move to the next logical level of methodological testing must mature to a stage where manned human missions can be attempted and brought in to significantly update this cycle of applied research. In missing this out, the deeper and clearer interrelationship of structure, energy, and how this is conceptualised, thus far, this first-hand perspective of evaluation pushes forward the science and the understanding that moves the entire field forward to create paradigm shifts in science, technology, as well as human self-insight. In working collaboratively to go to these new places it does utterly transform the technology to help us safely go to these new places – like exploring other planets and eventually other universes.

12

Conclusion

Why is space exploration important?

I have been interviewed many times about this subject of space exploration and have been a member of specialist professional panels looking at various aspects of this field. One of the surprisingly frequent questions that is asked is often, "Why do you think space travel is important?" There are many answers that can be given, but one that I think is genuinely valuable is that manned missions push the innovation envelope for technological spin-offs and applied science-based applications. Many of the biggest leaps forward with computing, AI (artificial intelligence), science, medicine, material science, and design innovation have been developed in large part through the requirements to problem-solve space exploration. What I think can be missed by outsiders are the multiple professions required to create the technological leaps and to problem solve to develop sustainable applications for these new technologies to properly work. What can be missed is how much ingenuity is actually required to rise to the next level of capacity.

There are several scientific and technical factors to exploring and colonising space:

> There are some critical technological barriers and scientific (medical) issues that must be overcome to be able to bridge where we currently are with our present capacity to embark on space exploration and where we need to get to make this dream an achievable reality. Some of these issues include:
>
> The impact of space environments and cosmological radiation on the human body.

The problems of muscle wasting and bone demineralisation (which greatly increase the risks of life-threatening bone fracturing and possible death).
The psychological impact of prolonged isolation.
The emotional dynamics of confined spaces and limited outside interaction.
The impact of potential chronic high stress and extreme physical danger.
Problems of sustainably and reliably growing food (for example through hydroponics) sufficient to sustain life within remote space bases.

All of these issues are possibly sufficient enough that one of them alone could halt a manageable plan to go beyond Earth's environment.

***Figure 12.1* NASA photo linking the smaller areas with a midpoint to show how vast the universe is**

Source: www.nasa.gov/images/content/731590main_Abell68-inset.jpg

These steps to explore outer space also bring with it significant potential worries, looking at commercialising asteroid and lunar mining. That there is the real potential prospect to cannibalise other environments and spread utilitarian environmental abuse, even beyond our own planet, seems like a looming possibility. One core principle related to psychological processes is how linked and interlocked, these psychical systems are involved in making sense of these relationships. The environmental destructive processes are example of how the predominate perspective can influence on how the environment is utilised. If these viewpoints go wrong it can vastly undermine our own ecosystem, as these are not only unresolved, but are also likely to be worsening. My point is that how these systems of dysfunction work is that they create stuck and enmeshed processes that alter the direction of healthy functioning (or on the other hand, create vulnerable processes). If we have not resolved some of the fundamental questions like unsustainable debt or educational problems, these earlier problems make later stages for advancement significantly more problematic. The risk if we have not sorted out our own environmental problems, then to move out to take these ecologically dysfunctional and unsustainable processes forward, catalyses a collective defensive fear, that from my understanding makes moving to these newer levels of evolution of space exploration basically impossible to reach if some thought is not put into addressing these factors. On a brighter note, if we do look into these aspects, they will also help mutually resolve a positive environmental stance on simultaneously moving to help innovate workable solutions to improve our own planetary health and practices, as well as to help make movements forward that can strengthen these best frameworks. As I began this short book, one of my deepest hopes is to help create a compassionate revolution for us to be mindful to completely and utterly transform our scope of what we feel that we can collaboratively work on accomplishing. Instead of destroying and hurting each other, what would happen if we could deeply work on these grand problems to completely change the quantum leaps of our collective shared inheritance of our accomplishments and our vision. Space exploration (and eventual colonisation) represents, from my analysis of all of these challenges, a critical puzzle piece that can snap into place the sharp focus that can put into greater order and proportional context meaningful answers that can show us the way forward.

References

Aschwanden, M. (2005) *Physics of the Solar Corona: An Introduction with Problems and Solutions*. New York: Springer Publications.

Beg, A. (2009) *The Importance of the Black Hole as an Organ of the Universe*. PDF Self published.

Bostrom, N. (2003) "http://www.simulation-argument.com/simulation.html" Are You Living in a Computer Simulation?. Philosophical Quarterly. 53 (211): 243–255.

Craig, A. D. (2009) How Do You Feel Now? The Anterior Insula and Human Awareness. *Nature Neuroscience Review*. 10, 59–70 pp.

Higgins, P. (2015) *Eradicating Ecocide*. New York: Shepheard-Walwyn Pub.

Low Earth Orbit (LEO) (2018). www.spacefoundation.org/programs/public-policy-and-government-affairs/introduction-space/orbits-earth

Maldacena, J. M. (1998) The Large N Limit of Super Conformal Field Theories and Super Gravity, hep-th/971120, *Advances in Theoretical and Mathematical Physics*. 2 (1998) 231–252 pp., *International Journal of Theoretical Physics*. 38 (1999) 1113–1133 pp.

NHS. (2009) www.nhs.uk/news/2009/01January/Pages/LowbloodoxygenlevelsonEverest.aspx

Susskind, L., and Lindsay, J. (2005) *An Introduction to Black Holes, Information and the String Theory Revolution: The Holographic Universe*. London: World Scientific.

Taylor Redd, N. (2012) Space.com (September 24, 2017), Earth-Sun Distance Measurement Redefined www.space.com/17733-earth-sun-distance-astronomical-unit.html

Telegraph. (2009) www.telegraph.co.uk/news/science/science-news/6660045/Bacteria-from-Mars-found-inside-ancient-meteorite.html

Uzan, J-P. (2010) Testing General Relativity: From Local to Cosmological Scales. *Philosophical Transactions of the Royal Society*. 369, 5042–5057 pp. DOI: 10.1098/rsta.2011.0293.

Vita-Finzi, C. (2016) *A History of the Solar System*. London: Springer Press.

Index

Note: Page numbers in italics indicate figures on the corresponding pages.